NEGATIVE IONS

NEGATIVE IONS

John Marino, the nationally-known cyclist, recently equipped his bicycle with a battery-powered air ionizer and set a world record for the cycle trip from California to New York.

Seymour M. Farber, M.D., former vice chancellor of the University of California Medical School in San Francisco, kept an air ionizer at his hospital bedside while recuperating from abdominal surgery during the spring of 1980 and declared that he healed faster as a result of being exposed to negative ions.

The State of California has installed ion generators to clean the air in its Los Angeles judicial hearing chambers.

The American Broadcasting Company has equipped its new thirty-story New York City headquarters with ion control through the use of generator boxes.

Two national appliance manufacturers, Philco and Emerson Electric, have spent thousands of dollars for research to create ion-control air-conditioning systems as part of their product lines. Radio Corporation of America, Westinghouse, General Electric, and Carrier Corporation have similar products in development stages and anticipate coming to the market with them soon.

A California chicken farm was ionized recently in an attempt to reduce brood hen mortality resulting from respiratory problems. The poultry farmer's effort succeeded beyond his expectations. The chickens, in fact, increased their egg production to a point where the entire negative ion generator system was paid for in four months merely from the difference in the number of eggs produced.

You, too, can personally benefit from using negative ions derived from ion generators which are readily available--frequently distributed by mail order or sold at holistic health conventions. The ionic boxes provide a variety of physiological advantages.

Clinical research on the positive effects of negative ions is scant and the findings are contested, but owners of the ion generators tell of relief from migraine headache, insomnia, asthma, depression, anxiety, joint pain, and a variety of other health problems. The few research scientists working in this area around the world, report benefits from incorporating negative ions in human surroundings, especially in building structures that substitute an essentially artificial environment for the real thing.

Investigators at the Environmental Protection Agency (EPA) have shown how particles charged by combination with air ions tend to attach to such grounded surfaces as ceilings, walls, and floors. This is deemed desirable because many airborne toxins, such as nitrates, sulfates, and lead from automobile emissions, are particles small enough to be inhaled if they are floating free in the atmosphere. It's the negative charge on the particles that make them react in a non-deleterious way toward human beings.

WHAT ARE THESE IONS IN THE AIR?

Ions are floating in the air around us all the time and have either negative or positive charges on them. Changes in their concentration, or in the ratio of positively to negatively charged molecules can have remarkable effects on plants and animals. It's known in science that ion depletion is the source of a wide range of human health problems, both mental and physical.

Air ion formation begins when enough energy acts on a gaseous molecule to eject radioactive substances in the Earth's crust, and some from cosmic rays. Electrons are displaced by this ejection. The displaced electron attaches itself to an adjacent molecule, which becomes a negative ion; the original molecule then becoming a positive ion. Repeated molecular collisions transfer the charge, so that positive charges come to reside on molecules with the lowest ionization potential, while electrons are attracted to the species of greatest stability. Next, neutral gas or water molecules cluster about the ions to form small air ions.

These air ions are important to you, because if they have a high proportion of negative ions in the clusters you'll feel lively, uplifted, and enthusiastic. Too many positive ions in the clusters will have you feeling depressed, lethargic and full of aches, pains and complaints.

Albert Krueger, M.D., LL.D., Emeritus Professor of bacteriology at the University of California, Berkeley, who is founder of the Air Ion Research Laboratory at that institution, points out that in normal clean air over land, there are 1500 to 4000 ions per cubic centimeter (cc). Negative ions are exceedingly mobile and the Earth's surface has a negative charge; therefore, negative ions are repelled from the Earth. This repulsion creates a normal ratio of positive to negative ions in the range of twelve to ten.[1]

---

1 Albert Kreuger. "Are negative ions good for you?" *New Scientist* 58:668-670, June 14, 1973.

If ions carry a negative charge, they are called negions; if they hold a positive charge, they're referred to as posions. Negions or posions are grouped into four types of "small ions" carrying their respective charges, and they look like this:

CO_2+ $H+$ O_2- $OH-$

Radioactive substances ejected in the Earth's crust or in action on gaseous molecules, as we mentioned, create negions by friction between wind and water. Following a storm there will be an abundance of negative ion because of the prior increased wind-water friction. Most negions are emitted by this form of "Earth generator" and pushed off various sharp points and edges by the "corona effect," which is a phenomenon in nature where a release of electricity similar to a flash of lightening takes place. The negions rise and are gradually neutralized by posions and water-vapor molecules they encounter. At the same time, posions are created by air-on-air friction at high wind lines and by the effects of solar radiation on the inosphere. Posions gradually descend to be neutralized by negions they encounter or by Earth itself. Most of the time they hold themselves in the balanced ratio described by Dr. Krueger of twelve posions to ten negions.

Negions are exceedingly beneficial for a person's metabolism as a means of enhancing human behavior. They act in a complex mechanism to bring about hormone and biochemical reactions in the body and brain. It is impossible to get an overdose of negions, which act like pure water in washing away dirty posions. The more negions you're exposed to, the better and more uplifted you feel.

Unfortunately, in a modern office or apartment building, the environment for negions is worsened as a result of our living in artificially constructed surroundings. Synthetic carpeting, plastic blinds or nylon drapes, office machinery, television or video displays, sealed windows, reinforced concrete, metal walls, fluorescent lighting, low acoustical tile ceilings, and air conditioning ducts and system elements all make for this artificially unhealthy environment.

Typical of the type of health problems arising from working in an artificial environment is the case of a microbiologist working for the Hawaii State Department of Agriculture in the Veterinary Laboratory. Corin Kanio became ill with a serotonin hyperfunction syndrome in September, 1980, after her department moved to a new laboratory-office complex in which there were no windows. The environment was automatically controlled by air conditioning and heat ducts. The microbiologist's work room was equipped with a negative pressure exhaust system, since she deals primarily with bacteria and fungi.

The serotonin hyperfunction syndrome or "irritation syndrome" involves sleeplessness, irritability, tension, migraine, nausea, heart palpitations, hot flushes with sweating or chills, tremor and dizziness. The elderly become depressed, apathetic and extremely fatigued. All these symptoms happened to Corin Kanio and she presently is undergoing grievance proceedings with the State of Hawaii concerning her work-related illness. She attributes it directly to the air ion sensitivity she has that comes from her work surroundings, which separate her from the air's negative ions and the Earth's positive field.

The following TABLE I compares the unhealthy atmosphere we build for ourselves indoors to the average number of negative ions per cubic centimeter in the great outdoors. And driving in an automobile with closed windows is even worse.

TABLE I

Average Ions Per cm^3

	Negions	Posions	Total
Clear mountain air	2,000	2,500	4,500
Normal land air	1,500	1,800	3,300
Before a storm	750	2,500	3,250
After a storm	2,500	750	3,250
Typical modern office	150	200	350
Closed moving vehicle	50	150	200

THE NEGATIVE EFFECTS OF POSIONS

Just as positive benefits are provided by immersing yourself in negions, negative effects come from breathing air with high density posions. As indicated in TABLE I, before a storm the posion concentration becomes more than three times the amount of negions. If the barometer is falling in anticipation of a storm, brace yourself! Adults, children, and animals alike react testily to such bad weather, for ahead is the kind of day that may leave you most prone to illness or accident, stupid mistakes, and irrational blowing off of steam. Posions are responsible. A little later we'll tell you why and how.

Doing some long-range personal forecasting will be revealing. The season of the year, the month you were born, the latitude in which you live are all factors that play a part in shaping not only your temperament but your physical and mental development as well. This won't be hocus pocus of any kind. What we are going to explain is based on verified scientific investigation.

Seven psychiatric studies disclose that babies born in the dreariest time of year between January and April incur a somewhat greater peril of later mental illness. Those born in the sunniest months from May through August seemingly are least disposed to psychiatric illness. Climate and day to day weather exert profound influences on the lives of these babies.

There is an impressive amount of evidence that connects low barometric pressure before storms with erratic behavior. This is a time where the posions in the air outnumber the negions by a ratio of over three to one. More suicide attempts take place, greater numbers and more serious accidents occur on the highway and in factories, and an elevated incidence of fainting spells seem to beset people. An analysis of the records of some 2000 public school students showed that their "conduct" marks sagged noticeably before a storm.

To balance air pressure when the barometer plummets, nature extracts air from the Earth together with traces of radioactive minerals. This strongly positive ion air encourages increased physical discomforts ranging from headache and nausea to the familiar rheumatic twinges that precede storms. Even the healing rate of wounds has been known to slow up, while the risk of infection accelerates. Tension and depression are certainly more common.

Conversely, negative ion air -- when the weather is quite comfortable -- not only stimulates morale but actually relieves certain chronic diseases. Experiments with negative ionization have shown marked improvement for sufferers from high blood pressure, asthma, and hay fever.[2]

Positive ionization contributes to the cracking of "fringe personalities," and sets off mentally ill minds. On summer dog days, when the posion index climbs, more neurotics slip across into pathological psychosis than at other times, psychiatrists agree. Sex crimes rise, as do all catagories of crimes of violence, and pilferage, bank holdups, and shoplifting become serious commercial problems.

Retail-store executives are more and more intrigued with the practical possiblility of reducing shoplifting by artificially controlling ion content of air in their stores. Air is ordinarily ionized by solar radiation, by nuclear radiation, or by being broken down in an electromagnetic field. Ionizing instruments, containing a radioisotope in an electrostatic field, remove positive ions from the air but leave the negative ions. Equipment would be installed as part of the store's air-conditioning system. We'll be describing these negative ion devices in the pages that follow. But first let's learn how posions and negions act on the body and brain to bring about physiological or metabolic effects.

2 H. Lee. "It's O.K. to blame the weather." *Pageant,* May 1962, p. 64.

HOW AIR IONS ACT ON THE WHOLE PERSON

Have you ever wondered why you feel irritable, jumpy, have aching corns, headache, and seem overly anxious during that time just before a storm? A lack of negative ions and an excess of positive ions are the culprits. The storm's onset produces an ion imbalance in the air before rain falls or snow settles down.

When relatively too many positive ions are present in the air before a storm, the positive charge is transferred in the air you breathe from your lungs to the blood, causing the blood platelets to release a hormone that quite strongly affects your moods, your joints, and other physiological functions in your body. The weather really does affect your routine. The following is a description of how air ions act on the whole person; you being that person:

During fair weather the atmospheric field causes charged air ions to migrate, producing a flow of current. Humans have evolved outdoors immersed in this electric current and scientists now believe that the current flow enhances various processes within our body, even to the cellular level. The same appears to be true even on the level of plants. People who work out of doors are at a distinct health advantage over those who are deskbound or labor inside factories.

The human body possesses a positive field, and a coupling or resonance between it and the Earth's field takes place. Such a resonance between systems brings about a transference of energy and perhaps accounts for some of the "charging" effects of being outdoors during fair weather.

Studies have shown that the natural oscillation frequency of the atmospheric field of seven to ten hertz (cycles per second) is the same as our brain's alpha state, our most relaxed yet alert state of being. This natural field oscillation, coupling with your own brain waves, enhances your alertness and improves your reaction time.

On the other hand, the three to five hertz waves produced by weather activity were shown to cause an increased reaction time. Slow reaction time may be dangerous under certain circumstances. This lengthened time that it takes for a driver to respond to an emergency situation, for example, produces a 31-41 percent increase in traffic accidents, quite measureable by the National Safety Council.

Much of the waste products of cell renewal are expelled through the skin. A positive electrostatic field around us draws these surplus waste ions away from the body, permitting rapid and unhampered renewal of all cells. This effect countributes to your general well-being.

In January 1979, for the gathering of firsthand information on the positive effects of negative ions, we visited an authority on biometerology, Professor Felix Gad Sulman, M.D., D.V.M., Chief of the Bioclimatology Unit, Department of Applied Pharmacology, School of Pharmacy Hebrew University - Hadassah Medical Center, Jerusalem, Israel, and his team of professionals in fifteen medical disciplines. Dr. Sulman explained how negative ions get into the blood stream to react on the body in the various ways we have already described and will elaborate on further.

Dr. Sulman, who is the discoverer of atmospheric negative ions and the original developer of the negative ion generator, said, "When oxygen is resorbed via the lung alveoles, the ionized molecules are taken up like the normal oxygen molecules and pass into the blood corpuscles.

The red blood corpuscles, whose hemoglobin is oxygenated by the air during inspiration, receive also the ionized oxygen. The negative charge is carried throughout the body, whereas the positive charge may attack the blood platelets (thrombocytes) which in sensitive patients release their irritating hormone, serotonin.

"Serotonin release produces dryness, burning and itching of the nose, nasal obstruction, headache, dry, scratchy throat, difficulty in swallowing, dry lips, dizziness, difficulty in breathing, and itching of the eyes. Dr. Krueger in Berkeley has shown that negative air ions do not produce this unpleasant serotonin syndrome; they act directly on the respiratory enzyme, cytochrome oxidase, and promote cell respiration," said Dr. Sulman.

All the tracheal changes attributed to positive air ions can be duplicated by the intravenous injection of serotonin. These effects can be reversed by treatment with negative air ions, he added.

"On the basis of these facts, it seems reasonable to postulate that positive air ions are 'serotonin releasers', and that a local accumulation of serotonin in the trachea is the immediate cause of positive ion effects.

"Negative air ions reverse positive ion effects by speeding up the rate at which serotonin is oxidized. Like other oxidase systems, monoamine oxidase is thought to consist of dehydrogenase linked to a respiratory chain which may include cytochromes or flavins. Krueger's experiments demonstrate that this same action may produce a cytochrome-linked oxidation of serotonin said the professor. "We have confirmed the 'Krueger Effect' in the testube. Positive ionization of blood increased its serotonin content in plasma, red blood cells, white blood cells, and thrombocytes. Negative ionization reduced the serotonin release significantly."

As Dr. Sulman suggested, with every breath, air ions are carried into your respiratory tract to be transferred across the lungs into the bloodstream where they are taken up by the thrombocytes. If they enter as part of molecules of oxygen, the negative ions or positive ions are transported by your red blood cells. Posions cause the thrombocytes to release the allergy producing hormone serotonin (5-HT). Serotonin acts as a bronchial constricter and slows the clearing action of the cilia thus allowing the mucus to thicken. If you are one of those many hayfever sufferers or have some other respiratory problem you'll find that serotonin greatly aggravates your conditions. There is less clearing of allergens and pollutants from your respiratory tract than from the lungs of others.

It is this powerful and versatile neurohormone, serotonin, which is responsible for many of the unpleasant symptoms exhibited by persons breathing positively ionized. air. In the words of Professor Sulman, "Some--particularly elderly people--may experience difficulty in breathing; asthmatics wheeze, rheumatic people feel their joints, and in general,sleeplessness or insomnia, irritability and tension are there for the asking. Hair and skin have an 'electric charge.' Migraine patients suffer with nausea and vomiting, and optical disturbances. Heart cases complain of palpitations, heart pain and oppression. Women before the age of menopause complain of hot flushes with sweats or chills. Hay fever patients get bad attacks of rhinitis with conjunctivitis, though this may not be the real hay fever season. Giddiness, tremor, and balance disturbances may appear, as well as diarrhea and a constant desire to urinate."

All this happens from stimulation of serotonin secretion by your being exposed to air containing high densities of positive ions. Since negative ions are able to reverse the effects of positive ions by speeding up the rate at which serotonin is oxidized by stimulating the action of monamine oxidase, it makes sense to stay in touch with quantities of negions wherever possible. As you'll soon learn, this is done with ease by employing negative ion generators.

In general, exposure to negatively ionized air has been shown to increase oxygenation of the lungs, vital capacity, and ciliary activity. All types of beneficial responses take place as a result of these friendly ions.

You experience a normalization of the activities of the endocrine glands, which plays an important role in reducing the effects of stress.

A shift occurs in your body's indicators of the blood acid-base balance towards alkalinity.

A sedating and pain relieving effect is experienced if you have undergone surgery or suffered burns.

Brain wave studies show an increase in brain amplitude with better thinking ability, a shift of alpha activity to the frontal area of the brain for better conceptual ability, and a higher synchronization of the right and left brain hemispheres for a more even balance of the personality.

Your body undertakes a normalizing of metabolism of vitamins and an increase in both static and dynamic work capacity and endurance.

If you're a lactating mother you find yourself able to produce more milk of higher quality for your infant.

Is it any wonder that all of us naturally seek out and enjoy particularly those natural areas of high negative ionization such as forests, mountains, beaches, and waterfalls? The healthy climate of mountains, the freshness of a pine forest, the exhilaration of air at the seashore, near falling water, or after a thunderstorm are all related to the electrical state of the air with its exceptionally high density of negions.

In pine or fir forests the needles on the trees, in conjunction with the atmospheric electric field, act as points for the generation of negative ions.

After thunderstorms, as the raindrops strike the earth, there is a splitting of the water droplets into larger positive particles and smaller particles containing the negative charge. This is known as the "Lenard effect." The smaller negative particles are repelled away from the negative surface of the Earth back up into the air.

At the beaches where there is rolling surf, splitting of water molecules through the Lenard effect once again comes in to produce an uplifted feeling.

The Lenard effect takes place around waterfalls, as well.

Of course, people visit spas on mountain tops, at beaches, and near waterfalls for many reasons, but the beneficial effects of negative air ions is likely to be one of the main ones, although not many know it.

IONIC WITCHES WINDS AROUND THE WORLD

In his scientific treatise, *Health, Weather and Climate,* Professor Felix Gad Sulman points out that "bioclimatology or biometeorology, the science of studying the influence of weather and climate on the living organism, is a young and at the same time a very old science. It is young by scientific standards and old if we consider the strong belief of man, from earliest historic times to the present day, in the great influence of weather and climate on a person's behavior and happiness and on the origin of his diseases."[3]

The Ancients often considered diseases to be due to possession by winds, and now we know from the investigations of Dr. Sulman and others that the ancients in some parts of the world were not much off the mark in their beliefs. We recognize, for example, certain witches winds in areas around the world that have people susceptible to them. Certain types of persons react with a series of allergic signs and symptoms. For these sensitive people, the old saying that "an ill wind blows no good" is all too true.

"Ill winds" or "evil winds" or "witches winds" really do blow in the United States and on other continents, as illustrated by the Santa Ana winds of lower California and winds along some stretches of desert from northern Arizona down into Mexico. The listing in TABLE II is almost the complete number of winds and their locations around the world that carry discomfort to human beings. They are dry, warm winds with a strong positive ion concentration that brings with them headaches and irritability, aggressiveness and listlessness. Wherever and whenever they blow, accident, suicide, and crime figures shoot up.

3. Felix Gad Sulman. *Health, Weather and Climate.* Basel, Switzerland: Karger, 1976, p.3.

TABLE II

Name of the Witches Wind	The Wind Location	Name of the Witches Wind	The Wind Location
Arizona Winds	Arizona, U.S.A.	Leveche	Morocco
Chinook	Northwest U.S.	Chergui	Algeria
Santa Ana	California, U.S.A.	Chili	Tunisia
Foehn	Germany and Switzerland	Ghibli	Libya and Tripolitania
Ikslok	Malta and Western Mediterranean	Shamal	Iraq
Khamsin	Arabian countries	Simum	Syria and Jordan
Levante	Spain	Simoon	North Africa
Mistral	France	Haine	Rumania
Sharav	Israel	Lips	Greece
Sharkieh	Egypt	Halmiak	Yugoslavia
Sirocco	Italy	Pyrnwind	Hungary
Thar	India	Northern Winds	Melbourne, Australia
Zonda	Argentina	Canterbury	New Zealand

"The phenomenon that brings on one of these witches winds consists of a warm upper-air mass which descends along the slope of a colder air mass and manifests itself as a ground level wind, often carrying dust and debris from distant areas of its cyclonic origin," explained Charles Wallach, a behavioral science consultant and researcher in brain physiology, who is president of the International Bio-Environmental Foundation of Silver Springs, Maryland. "During its passage aloft such a wind will generate an electrostatic field by friction with the higher and lower air masses between which it is passing. This causes the wind to attract and retain posions. A 'posion rain' can be sensed through the thinning lower layer of air long before the warm air mass has descended to ground level."

The warm and dry electrified atmosphere exerts a strain on your body, to which it responds, in accordance with Selye's theory of stress, with an alarm reaction. The adrenal medulla produces increased adrenaline whereby your blood vessels, which had become dilated as an initial reaction to the heat, again become constricted and perspiration is reduced. Your adrenals become depleted when exposed to the witches wind and you get exhausted, even with doing nothing. Additionally,thyroid hyperactivity provoked by this hot, dry wind and its pressure variations hit. These come upon you while you are suffering with a serotonin release by the positive electrical charges occurring one or two days before the arrival of the witches wind.

Obese persons suffer more from these internal reactions because obesity is conducive to heat accumulation and reduced heat dissipation.

As described, these witches winds are accompanied by peculiar biological effects in the residents of the countryside where the wind strikes. Thus, whenever the descending wind exists, its electrical impact on the population is feared and hated. If these winds are cold, they are less dreaded though rather unpleasant. But all of them bring side effects that pervasively invade the life of everyone in their paths. That the witches wind blows no good is all to true.

WE CREATE ARTIFICIAL WITCHES WINDS

Everywhere in the industrialized western countries where a natural witches wind does not exist but modern structures of steel, concrete, and plastic are erected, we create our personal artificial witches winds. Western technology is distorting the electrical charge of the air where people spend their time. Office buildings with central heating and air conditioning systems, hermetically sealed high-rise apartments, automobiles, airplanes, synthetic fibers used for clothing and home furnishings, smoking in public places and private homes, city pollution, industrial wastes, and all the other ecological changes that make for modern living are causing us to live constantly in a positive ion-poisoned atmosphere. It's as if the witches winds that blow only occasionally out of the Arabian desert are with us constantly.

Fred Soyka and Alan Edmonds in their book *The Ion Effect* report that more than 5,000 scientific papers in multiple languages point up the conditions for the prevalence of witches winds, and most of those conditions we build for ourselves right within our urban environments.[4]

Office and factory workers, clerks in stores and shipping depots, and drivers of vehicles, pilots, and other flying personnel don't fare too well with their indoor environments. The structure of their building or vehicle serves to shield out the natural electric field, cutting them off from its physiological benefits. Americans are exposed to insufficient negative ions. Without the negative ions natural cleaning effect, the air indoors becomes loaded with pollutants and germs. Workers no longer experience an energy coupling with the natural electric field and are left to their own physical reserves. Negions' counteracting effects on the atmospheric pressure is lost to indoor workers and they become more sensitive to the weather and its posions.

Most modern structures are designed to be energy efficient and so depend upon air circulation from a mechanical ventilating system. This has led to a stripping of the negative ions from the air and a marked increase in positive ions. Illness among personnel runs rampant without the workers or the employers realizing that employee absences are directly due to the poor atmospheric status of the workplace. Most indoor environments measure out to have either an imbalance in the ratio of positive to negative ions or a depleted level of negative ions.

Is it any wonder then that so many personnel in modern office buildings are complaining of headaches, drowsiness, nausea, lack of ability to concentrate, and a general washed out feeling? Persons who are susceptible to these polluted environments, the elderly and young, and those with health problems find that this unnatural elec-

4. Fred Soyka and Alan Edmonds. *The Ion Effect.* New York: Bantam Books, Inc., 1978.

trical state of their surroundings greatly contributes to the seriousness and degree of aggravation of their condition.

For example, a mysterious illness had been plaguing employees of a medical-services company, the Itel Corporation of Port Washington, Long Island, New York, for six months. Federal, state and Nassau County public health officials could say only that they believed it was "something in the air." Despite numerous inspections of the new one-story brick building in which the company was lodged and examinations of its 300 employees who had become ill, the medical detectives were wtill unable to pinpoint the source of trouble. The personnel health problems started with the report of "a musty odor."

Since that first report March 20, 1979, the company's employees, mostly women, developed such symptoms as nausea, drowsiness, puffy eyes, lightheadedness, and general malaise. A check of the new air-conditioning system was made and it was found to be working well -- too well. Unknown to the officials, large quantities of posions were overshadowing any possible negative ions prevailing in the office building atmosphere. Four pregnant women were told not to report to work until the cause of the illnesses was found. The company hired investigators and a physician to find the source of the troublc, and they were kept on the payroll.

It cost the company about $50,000 in expenditures and another $250,00 in lost production, but finally it was discovered that positive ion replacement with more negions would solve the problem. Before this simple solution was accepted; however, all kinds of other causes had to be ruled out.

Philip R. Taylor, M.D. of the United States Center for Disease Control in Atlanta was assigned to the New York State Health Department to investigate the Itel Corporation building conditions. Dr. Taylor correctly thought the problem involved the air inside the building. He requested that a questionnaire be filled out by the company's group in efforts to determine whether there was a common denominator involved in the illnesses.

Thomas Abruzzo, M.D., director of the County Health Department's Bureau of Disease Control, at first attributed the illnesses to an unknown virus. Six months later he said that the illnesses were not virus infections. Eventually two problems were decided upon. Small amounts of some kind of irritating compound such as formaldehyde, which is used in glues, plastics and furniture was suspected and the air circulation within the building was pin-pointed. The air loaded with posions came from the overly efficient air conditioning system which excludes fresh, clean air from the outside. At times, the negative ion content dropped to almost zero. "The ultimate answer to what has been causing illness," Dr. Abruzzo said, "could have a big effect, from redesigning buildings to eliminate certain chemicals."[5]

How buildings may be readjusted to overcome the unhealthy effects of insufficient negative ions in the local atmosphere will be discussed in detail in the next section.

RESTORING NEGATIVE IONS TO YOUR AIR

How does this information affect your health and production? Fortunately, through dedicated research and advances in modern technology, we are finally able to control the electrical state of our indoor environments. The technique involves the use of systems to recreate the natural electrostatic field, with its normal Schumann resonance (air wave frequency), and to generate negative ions back into the air. The simplicity of the technology's application and the economics of its use bring us closer to the time when the interiors of our buildings will be returned to a more natural and beneficial state for the persons living, working, playing, eating, and sleeping inside. We will eventually be enjoying fair weather and comfortable surroundings indoors by application of ionotherapy.

Ionotherapy has been widely introduced in Europe in recent years and rightly so. It can do no harm, yet it can help the weather-sensitive population. No wonder, therefore, that recently everything in Europe has become ionized: people, bedrooms, offices, houses, factories, elevators,

5. Roy R. Silver, "Odd illness idling staff at L.I. plant." *The New York Times* , September 12, 1979, p.B2

telephone booths, cars and also cold storage plants and refrigerators. The only question to be pondered is when will the popularity of ionotherapy reach the United States?

There are new apparatuses which emit negative ions in order to neutralize the noxious effects of positive ions. The modern ionizing apparatus is called negative ion generators.

Dr. Sulman said that not everybody enjoys negative ionization -- only 30 percent of the population really need it and another 30 percent who feel well if the amount of negions in the air is increased. About 40 percent don't notice negative ionization as a positive benefit.

Noxious effects from use of a negative ion generator have never been described, because negative ions decay within a few seconds. The negions can't be preserved. But the generators do a fine job of producing negions: 50 to 500,000 negative ions per cc at a distance of one meter. At two meters, the negions fluctuate between 20 to 100,000 per cc. The optimal range for a person wanting to reap benefit from the negative ion generator is two to four meters distance from the device. One meter equals 3.3 feet.

THE THERAPEUTIC USES OF NEGATIVE IONS

Therapeutic use of negative ions requires no special arrangement. You may switch on the generator at night or labor with it in front of you on your desk or at your work bench. Keeping the windows closed is unnecessary, and running the generator cojointly with humidifiers or air cleaners is also unnecessary, although desirable. A combination with air conditioning is permissible if there is ion control on the air conditioner so as to not affect the ionization. Place your negative ion generator at the exit of the air conditioning duct.

Robert Massy, Ph.D., Chairman of the Physics Department of University of the Trees and Director of the Ion Research Center, Palo Alto, California, in an interview described some of the therapeutic uses of negative ions.

"People who are mainly involved with physical activities have reported an increase in physical energy levels and a reduction in sleep," said Dr. Massy. "Intellectual people, who tend to work mainly with their mind, have reported that they do not experience mental fatigue while the machine is on, and find that their thinking is sharper.

"The results of people who are feeling insecure, leading to insomnia, experience a decrease in anxiety and an increase in the number of hours slept. This second result is totally the opposite to the physically oriented person. The negative manifestation of those who operate mainly from the conceptual way of thinking is depression. We have had reports of depression being lifted when in the presence of a negative ion generator," Massy continued. "Imaginative people state that their dreams become more vivid and are in brilliant technicolor. People who have suffered from physical ailments have reported relief, but it is not wise to draw any conclusions until much more data have been collected."[6]

Igho H. Kornblueh, M.D. of Philadelphia (now deceased) demonstrated that negions had healing qualities for burns and postoperative wounds.[7]

Also, patients with respiratory problems such as chronic bronchitis or asthma report that ionotherapy helps them to breathe more easily. Positive ions make breathing more difficult. Other respiratory conditions benefiting from negative ion therapy are pulmonary emphysema, laryngitis, pharynigitis, dry hacking cough, infections of the upper respiratory tract, and nasal and sinus conditions caused by allergies.

Harry L. Bechman, M.D. of Bedford, New York uses what he calls "oxygen ion therapy" for patients with bedsores. Negions and oxygen gas are packed into plastic bags placed over the bedsores; healing takes place swiftly, faster than the ordinary time span for a bedsore to close.

Peter Fox, M.D. of Dorchester, England applies negion therapy for the relief of migraine headaches.

---

6. Susan Welker. "Are negative ions necessary?" *The Herbalist,* April/May, 1980, pp. 21 & 22.
7. I. H. Kornblueh and T.A. David. *American Journal of Physical Medicine,* volume 39 no. 3, June, 1960.

Other health difficulties reported in the medical literature as being benefited by negative ion therapy are cancer, sinusitis, joint disease, high blood pressure, pulmonary tuberculosis, and diseases of the peripheral nervous system such as peripheral neuritis.

Dr. Albert Krueger's research at the School of Public Health, University of California, Berkeley shows that doses of positive ions make mice more susceptible to disease. Negative ions cut back the production of serotonin, which had an effect similar to giving a tranquilizer -- mellowing the mood of animals and people. Kruegar and his associates analyzed more than 12,000 mouse brain samples and 36,000 blood samples from ion-treated mice. The mice and rats dosed with negative ions became more relaxed in stressful situations.

In another experiment, mice were injected with influenza germs. Some were given strong doses of negative ions and others were not. Twice as many of the ionized mice survived as the untreated. Kruegar said this shows that negative ions help kill bacteria and viruses.

There are several ways to avoid positive ions and get more negative ions. First, whenever you are able, wear natural clothing fibers such as cotton, linen, wool or silk. Avoid the synthetic products, since they promote the storing of static electricity.

Then, furnish your home and workplace with items made of more natural building and decorating materials. Have your furniture made of wood rather than plastic or metal. Lay wool carpeting on the floor. Natural fabrics should hang as draperies and curtains.

Live in a house made of wood, stone, stucco, bricks or other building materials usually found in nature.

Stay away from artificially climate-controlled environments at home or at work such as air conditioning. Another thing, surround yourself with growing green plants. Thriving plants provide a direct illustration that you are living in a ion-enhanced atmosphere, especially if they havc a marked increase in size and growth rate, and they tend to gather negative ions around themselves. This is particularly true of evergreens and ferns, which focus negative electrical charges on the pointed tips of their leaves.[8]

Finally, drink eight glasses of water per day that doctors recommend, shower often, walk barefoot in the dew in the early morning, control the movements of ions in your own living space by avoiding surfaces which tend to develop a negative static charge such as the soles of much modern footwear, and employ modern technology to create an environment similar to nature's healthier examples. Accomplish this by installing negative ion generators in your home and place of work.

The ion generator you acquire should produce negative ions by means of the "high voltage corona discharge effect." There are many brands of generators on the market that produce negative ions, and the one you use should contain a transformer which converts the 110 volts of local power to the high energy 5000 volts required to force 62.5 billion electrons through each of the generator's needles you find in the apertures or the metal mesh that acts as a total electron outputter. About 250 billion electrons per second is sufficient for the optimal negative ion effect.

During every micro-second of operation these discharging electrons simultaneously attach themselves to oxygen molecules. This results in each of the oxygen molecules receiving an extra electron, making a negion out of the negatively charged oxygen.

Your machine should carry a guarantee for a year against defective parts or labor. It must be double grounded for safety. You should be able to merely plug it into your wall socket and turn on the switch. Make sure the generator's emitter needles are stainless or nickel-plated steel and not an alloy that corrodes with time; otherwise, your machine will become useless.

For more information, contact the Bio-Environmental International Foundation, the research organization on negative ion generators. It's located at 1220 Blair Mill Road, Suite 1205, Silver Springs, Maryland 20910; (301) 585-8881.

---

8. Morton Walker. "Are you breathing malnourished air? *Prevention,* July 1979, pp. 93-98.